Maintenance Manager's Guide to Work Management

Administering Asset Management Plans

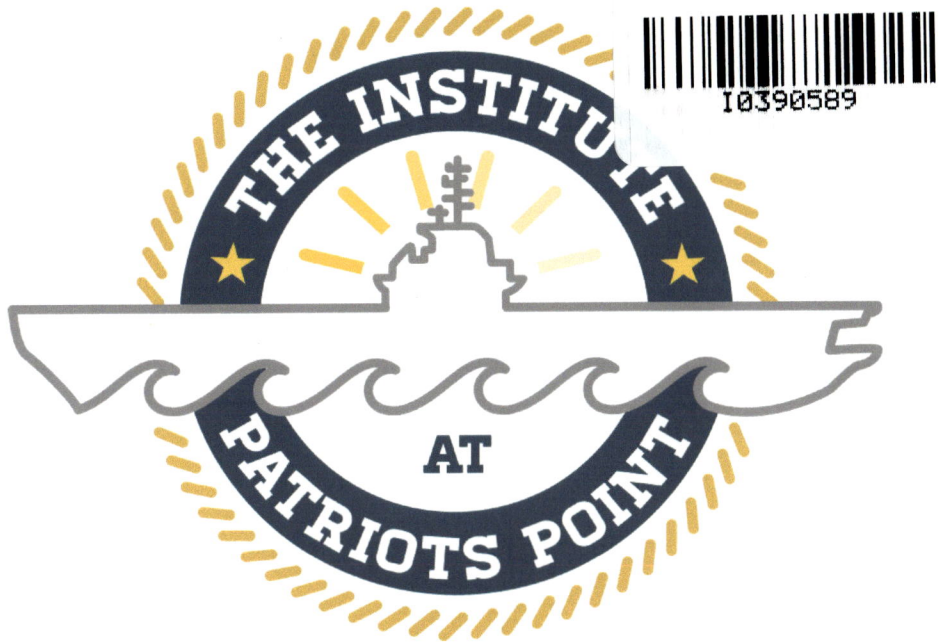

Authored by

Darrin J. Wikoff
Brandon Weil

TABLE OF CONTENTS

LIST OF TABLES

the way work is performed today, while identifying opportunities to improve the process to meet tomorrow's needs and expectations.

Once the "As-Is" process is defined, best practices can be introduced as a comparison or "Should-Be" state. In most cases, the gap between the "As-Is" and "Should-Be" is too great to immediately improve process performance and, if incremental improvements are not made, often creates greater inefficiencies. For this reason, the "To-Be" or target state process is mapped to bridge the gap. The "To-Be" process allows the organization to continually improve process performance by removing human inefficiencies, optimizing the process through technological advancements, all the while striving to reach the "Should-Be" or ideal state.

Work control processes also provide a means by which organizational performance can be measured for continuous improvement. Because the work control processes are developed as the standard for performing work, and providing that the organization is disciplined to consistently execute the process as defined, Key Performance Indicators (KPI) should be developed to identify inherent inefficiencies within, and non-compliance to the process. Indicators such as Schedule Compliance, Stock Outs, and Overall Equipment Effectiveness (OEE) measure the organizations ability to adhere to and collectively execute each work control process.

Work control processes are fundamental to improving overall organizational performance. Each process ensures standardization of work, providing consistent results to meet the desired expectations. Work control provides the organization with the ability to recognize ineffective processes and quickly eliminate inefficiencies that result in increased operating costs. Formal documentation of work control processes and focused process improvement ensures sustainable growth and higher productivity.

Principles of Asset Management

Before we launch into a more detail discussion about the Maintenance work control process and associated business systems, let's begin by defining and understanding the core principles associated with asset management. To keep this conversation practical, we'll use an automobile as a familiar reference to explain how each principle applies to your business systems.

Assets

First, let's define the term "Asset". An asset, with regards to capital, is any item that has a quantifiable value to the business and its stakeholders. Relative to the automobile, value is quantifiable in terms of equity based on the purchase price or resale value, and the loan amount, as a liability, secured to purchase the vehicle. The automobile has a value, but we might also divide the vehicle into smaller groupings of assets that have a uniquely different value. Tires, for example, as a replaceable item will require additional investment from the vehicle's user over the life of the parent asset, the automobile. As such, the tires may also be an "asset", either as a grouped system of components or as individual items. Because the tires can be used over more than one fiscal or financial reporting period they are a capitalized item, an "asset". The cost of replacing the tires serves as the initial value needed to define the asset. We can expand the definition of an asset in the context of reliability by adding the fact that an asset must also have a quantifiably distinct function or purpose within the business. Remember when we said that the tires associated with the automobile could be an asset as a grouping of components or as individual items? This definition explains why.

If we assume that the function or purpose of the tires, in the context of the parent asset, is to provide friction between the drive train and the road to achieve 85% power transmission efficiency, then the asset formerly known as "tires" should be divided into a minimum of two unique assets. In terms of the power transmission efficiency function, the two rear tires, on a rear-wheel drive automobile, would be classified as a single asset. Because you need both rear tires to create the required level of power transmission efficiency, the asset is "Rear Tires". This grouping of components within the asset definition is known as a "System". The system is the unique asset and a single asset identification number would be assigned to the "Rear Tires". Because the front tires do not enhance or impede the power transmission efficiency function, they are not part of the "Rear Tires" asset. As such, the front tires must have a different purpose or function as a system or as individual assets and would be assigned an asset identification number different than the "Rear Tires".

Risk

Once we understand what assets exist within our business, and we have defined the quantifiable values and functions of each asset, we shift our attention towards identifying the risks associated with each asset's ability to transform capital into money or cash. This is, after all, the reason why we made the investment in the first place. Managing the risks associated with physical assets requires a focus in two parallel areas: Effect – the consequence associated with the asset *not* performing its required function as desired – and Probability – the level of uncertainty that the asset *will* perform its required function as desired.

The easier of the two focus areas to begin with is the discussion regarding the *Effect* a non-performing asset has on the business. In the setting of the automobile, the most significant effect would be based on your inability to drive the car to and from your place of employment as the user or primary stakeholder. Quantifying the *Effect*, in this case, would be determined by the amount of time you are unable to drive the vehicle and the income you lost within this period of time. Risk, in these terms, is measured as:

$$Effect = \frac{\text{Income}}{\text{Time}}$$

However, since you are not the only stakeholder associated with the value of the automobile, we must also consider other risks. If you secured a loan to purchase the vehicle, then the Bank may be a stakeholder. The Bank's risk is based on you not being able to make your loan payment on time while you are unable to use the vehicle to generate an income. The interest lost on the Bank's asset would replace your income in the numerator of the risk equation. You might also consider the county's stake in the automobile asset, or the impact the unusable vehicle might have on fuel sales in your local community as another stakeholder. In general, when quantifying the *Effect* risk associated with an asset you must consider the following:

- What is the impact to the primary stakeholder or user?
- What is the impact to other financial stakeholders?
- What is the impact to the physical environment in which the asset operates?
- What is the impact to other assets in the value stream?

The "Probability" risk associated with an asset refers to the level of uncertainty that an asset will perform its function as desired. *Probability* is a measure of confidence in your ability to ensure value – convert the capital investment into cash. Uncertainty because of not knowing what will go wrong or what may cause the asset to fail to meet stakeholder expectations is the risk we are trying to quantify. *Probability* weights the *Effect* to justify technical and financial decisions.

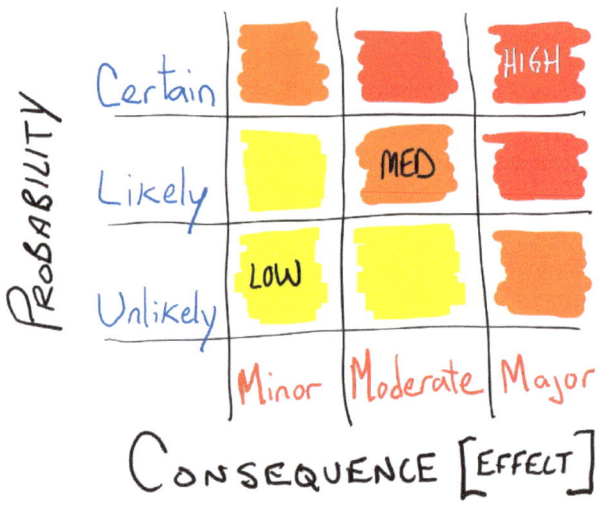

Figure 1 - Risk Matrix

The asset Risk Matrix illustrates the relationship between *Probability* and *Effect*. In this model, the consequence associated with the asset underperforming is the constant and assumes that the *Effect* does not change as long as the function or purpose of the asset remains the same. This is important to note when evaluating the risks associated with existing assets. "Has the function or purpose of the asset changed?" In many cases, as a company grows in market share or launches a new product, the function of core assets within the company changes. With the new function comes new risks. *Probability*, in comparison, is variable based on the operating context associated with how the asset is being used. We must also keep in mind that *Probability* equates to the level of uncertainty surrounding the asset. Returning to our automobile analogy to help us understand these core principles, would the level of uncertainty increase if you changed how you operated the vehicle? Assume your normal operating mode is to drive 20 miles from home to work and back again five days every week. How certain are you that the vehicle will get you from point 'A' to point 'B' without breaking down on the side of the road? Pretty confident I'm sure. You make sure each day that you have enough fuel to make the trip, and occasionally have your vehicle serviced to ensure it is in good working condition. Now, assume that you are taking a 1,400-mile road trip from Dallas, Texas to Los Angeles, California across the deserts of Arizona and through the steep mountain passes in eastern California. Would you be as confident in your preparations and current asset management practices? Would you decide to take extra precautions to ensure your 1,400-mile trip was successful? Sure you would! You might have the tires rotated, balanced and inflated to a lesser pressure to compensate for the extreme climates. You might even purchase a new spare tire and extra fuel tank knowing that the 200-mile journey through the desert is without frequent service stations. What about you as the Operator of the automobile. Would you change your driving practices too? You might set a cruising speed of 65 mph to optimize fuel economy or make frequent stops to check your tires and other mechanical components that are subject to fatigue in the extreme climates. All of these technical and financial decisions are based on the increasing level of uncertainty resulting from the new operating mode. The function of the asset hasn't change. The desire is to travel from point 'A' to point 'B' without breaking down. The consequences haven't changed. Any breakdown of the automobile will result in the same *Effect*, loss of income, increase in expenses or loss of usability in general. The only risk parameter that has changed based on the operating context is the level of uncertainty.

Asset Management Plans

If assets exist within a business to increase the business' ability to generate revenue, then our asset management plans and activities should center on mitigating the risk to said value. The last asset management principle we are going to discuss in this introduction deals with the approach to asset management planning. Asset management plans define how a business ensures value. The genesis of each plan to operate or maintain is derived from the identified and quantified level of risk deemed unacceptable by stakeholders. Asset management plans must clearly link the identified business risks to prescribed activities or tasks and the resources required to accomplish each activity. These plans are not limited to countermeasures as reactionary devices used to recover once an undesirable effect has occurred. Instead, these plans are focused on the prevention of risk all together. In doing so, preventing the risk, the level of uncertainty diminishes and the overall risk to the business is reduced.

Asset management planning requires a systematic approach to ensure that all feasible causes and effects are identified, and that the business has provisioned for those activities that will mitigate the risk to a suitable level. Asset management planning typically follows the following steps:

1. Risk Analysis – Evaluating each operating mode to determine the causes of undesirable effects.
2. Task Selection – Identifying the operating or maintenance practice that will prevent the undesirable effect, or in the least predict the conditions that will result in the undesirable effect if unmanaged.
3. Resource Planning – Determining the skills, competencies and informational requirements necessary to consistently execute the prescribed tasks.
4. Budgeting – Allocating the financial resources required by the resource plans.

Asset management plans are often closely tied to functional departments and operating budgets through a formal, and more strategic business plan. Although these plans exist under many different titles, the objective is the same – manage the risk. Managing the risk boils down to one of two choices: *Accept* the risk or *Defer* the risk. When facilitating the asset management planning process, it is important to understand the risk threshold for each stakeholder. Threshold, in this discussion, means the level of risk that the stakeholder believes is not worth managing. Our assets represent the capital invested by the company in the future of the business. If the cost of managing the risk outweighs the future value inherent within our assets, then we may *defer* the risk. Deferring the risk doesn't mean pushing it off onto someone else. To *defer* means to postpone the activities to another time period. A good example would be, coming back to the automobile application, to *defer* the replacement of all four tires until after you have finished paying off your loan. By eliminating the loan liability you are increasing your equity in the asset, and now the capital expenditure needed to improve the reliability of the tires is justifiable. Additionally, if the level of risk – Effect and Probability – associated with one asset is insignificant relative to other assets, it may not be economical to allocate money to the activities and tasks prescribed through the management plans. In these cases it is better to *accept* the risk and implement countermeasures that assist the organization with recovering lost value short-term.

Again, regarding the automobile, there is considerable information and data supporting the idea that the likelihood of a catastrophic engine failure in the first 35,000 miles is low, although the consequence could be high. Using the asset Risk Matrix, the "engine failure" mode might be less significant than the risk associated with a mode that we are less certain about. In order to mitigate the risk without incurring extraordinary costs – like overhauling the engine every 10,000 miles, the financial and technical decision would be to *accept* the risk and purchase a 35,000-mile warranty service plan to assist with consequence recovery should an engine failure occur. With both decisions to either *accept* or *defer* the risk, the goal is to maximize the future value of the asset based on the quantified level of risk short-term.

When designing, administering or evaluating your maintenance work management business systems, keep these asset management principles in mind. Maintenance must be capable of managing varying levels of business risk. The maintenance business systems must not add to the risk profile of the business, and, resources and budgets must be allocated appropriately to effectively manage business risks without increasing the total cost of asset ownership beyond an acceptable level.

THE MAINTENANCE WORK CONTROL PROCESS

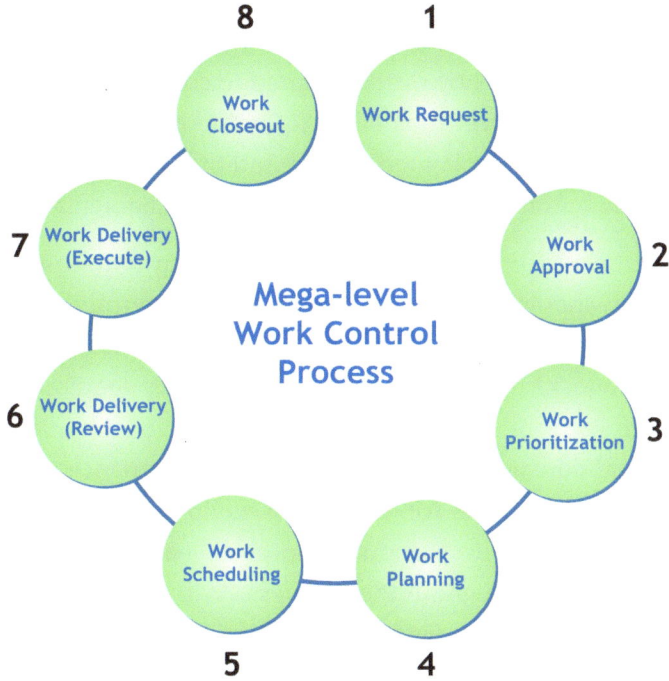

Figure 2 - The 8 Operations of Maintenance Work Control

At the mega-level, the maintenance work control process follows a basic eight-step sequence of operations that begins with the initiation of work and ends when the work order is "closed", providing the data required for asset and work history reviews performed through reliability management processes. We will discuss the best practices within each operation of the maintenance work control process that ensure assets are maintained in an efficient and effective manner.

Work Request

The maintenance work control mega-level process begins with a formal request for work to be accomplished based on the need for production support, the need to recover from a breakdown, safety, or environmental incident, or based on a suggestion to improve the condition of an existing asset as found through audits, preventive or predictive maintenance inspections, or condition monitoring reports.

The work request is also used to identify the need for maintenance resource support for capital or engineering projects as part of the maintenance backlog.

The work request *is not*, however, designed to communicate or initiate a preventive or predictive maintenance work order, as these work classifications should be planned, forecasted based on a predetermined frequency, and automatically "dropped" into the schedule as defined.

To improve the efficiency of maintenance, or mean-time-to-repair, each work request should be screened against predetermined criteria to ensure the appropriate level of detail is provided to accurately identify the scope-of-work, asset for which the work is required, and the requestor to provide a means for follow-up communication and feedback. Additionally, as the output of the maintenance work control process is to provide data for evaluating equipment and work history records, the work request should allow for easy identification and trending of failures throughout the business. Within most Computerized Maintenance Management Systems (CMMS), a Failure Code can be used to facilitate this aspect of work identification to make trending, reporting, and analysis of equipment reliability more efficient and timely.

Level 1	Level 2	Level 3	Level 4	Level 5
Work Requests are not used; Work reporting is informal.	A formal system for reporting work exists; Verbal requests are predominant.	Work is requested in CMMS; Limited access to system; Most requests are post-work completion.	> 80% of work is requested; Work request clearly identifies scope.	Duplicate work requests are minimized; Failure coding enables trend analysis.

Work Approval

The next step in the maintenance work control mega-level process is a formal procedure for approving work to ensure maintenance and engineering expenditures are supportive of the overall business needs, eliminating the redundant, "wish-list" and less critical tasks that can bog down the backlog, providing an inaccurate illustration of the current labor resource requirements.

The approval process should be performed by a single-point accountable role that *has the authority to make decisions without the consultation of other departments*, accept when clarification is required. The process should be facilitated by a predefined set of criteria, and an Approval Code which provides a means for coding work requests that are not approved as a form of feedback to the originator. The matrix may also serve as a training tool to improve the integrity of work requests that are frequently refused due to lack of information.

Because of the approval process, the work request should be transitioned to an "Approved" work order, or coded and returned to the originator for additional information and removed from the system.

A common problem when trying to facilitate work approval is not being able to remove the emotional and social aspects of the decision-making process. Larger organizations which have political régimes imbedded in the culture are often unsuccessful without the help of a decision or logic tree. In this case, work approval should be combined with the third step in the mega-level process, entitled Work Prioritization.

Critical Success Factors:

- The work approval process is formalized and supported by an Approval Codes Matrix that serves as a means for providing feedback to the originator, and as a training tool for improving work request integrity.

- The work approval process provides a means to prevent the duplication of work requests, and subsequent work orders, that have the potential to falsely inflate the maintenance "backlog" – all approved work not completed.

- The approval process is automated when possible to reduce the lead time from Work Request to Work Order.

Work Prioritization

Level 1	Level 2	Level 3	Level 4	Level 5
Operating areas receive higher priority; Priorities are subjective.	Work orders are prioritized by Maintenance Supervisor; Operations schedule dictates priority; No risk method applied.	Work orders are prioritized by either asset criticality, defect severity or work order type; Formal system documented, but not consistently applied.	Work orders prioritized by both asset criticality and defect severity; Formal system documented and consistently followed.	Level 4 and includes a work order aging factor to increase risk priority; Priority trends are used to improve methodology.

In support of the work approval process, the work order must be prioritized within the backlog to guide Production, Maintenance, Engineering, and Materials management through their daily activities of meeting equipment needs. The work prioritization methodology must be structured to accurately differentiate between the emergent, urgent, essential, desirable, and routine work. The prioritization model must be robust enough to ensure effective allocation of maintenance and engineering expenditures. Work prioritization should also consider the impact an asset has on the overall business performance, as well as, the risk of not completing the identified work within acceptable lead-times.

As a rule, work prioritization is fundamental and can often be informal, however, a more rigid and formalized process can greatly improve the organizations ability to control maintenance and engineering expenditures based on the business needs. Within most industries, as a minimum standard, work prioritization includes 4 to 5 scheduling horizons as shown in Figure 3.

Schedule Guide	
1	Must be <u>completed</u> immediately using all required resources.
2	Must be completed within 24 hours (1 day) using available resources as required.
3	Must be completed within 2 to 7 days using predetermined resources as available.
4	Must be completed within 8 to 30 days as scheduled based on resource availability.
5	Must be completed within 31 to 42 days based on scheduled resource availability.

Figure 3 - Scheduling Horizon Priorities

As this method provides an acceptable guide to managing schedule compliance, it *does not* provide enough delineation within each priority code to accurately manage the allocation of maintenance and engineering expenditures based on those activities which will have the greatest impact on the business. As an organization becomes more proactive, the greater percentages of work tend to fall into the priority '4' and '5' categories. For this reason, many organizations today have moved towards the RIME model of prioritizing work.

Ranking Index for Maintenance Expenditures (RIME)

The method for obtaining the RIME priority values is one that is reasonably objective and produces a quantitative index of the relative importance. The equipment to be worked on (Equipment Code) and the type of work to be performed (Work Order Class) at the time the need for the job occurs.

The RIME priority of a work order is determined by establishing a numerical value that is the product of multiplying the "Equipment Code" by the "Work Order Class" for the work and equipment or facility involved on the request. The highest index number is the most important job, and the lowest index number is the least important job.

RIME Equipment Code

Each piece of equipment, machine, and building is placed in one of nine (9) Equipment Code categories. The most important items carry a code value of nine (9), and the least important have a code value of one (1).

[9] UTILITIES
Primary utility equipment that may cause an unplanned outage of one or more areas of the plant.

[8] PROCESS EQUIPMENT (Not spared)
Equipment necessary for processing the primary product within the value stream. The equipment cannot be bypassed by manual means or by utilizing additional manpower. Failures will stop the production process.

[7] PROCESS EQUIPMENT (Spared)
Equipment necessary for production but redundant equipment or bypass method is available. Failures will not influence the production process.

[6] SUPPORT EQUIPMENT (Not Spared)
Equipment used in an auxiliary capacity to the production process (e.g. hot water recovery pumps, auxiliary high-pressure compressor, etc.).

[5] SUPPORT EQUIPMENT (Spared)
Equipment used in an auxiliary capacity to the production process, but redundant equipment is available.

[4] MATERIAL HANDLING EQUIPMENT
All equipment associated with the movement of product or raw materials (e.g. fork trucks, cranes, hand trucks, etc.).

[3] PRODUCTION FACILITIES, OFFICIES, LABS, SHOPS
Physical facilities that serve production but do not have a direct impact on production.

[2] PRODUCTION EQUIPMENT LAYDOWN

Any stored equipment that will be out of service for 30 days or more.

[1] BUILDING FACILITIES AND GROUNDS

Office areas and grounds that are not used to support production.

RIME Work Order Class

All work performed by the Maintenance Department is separated into nine (9) Work Order Classes. The most important work is Class nine (9) and the least important work is Class one (1).

[9] SAFETY, BREAKDOWN, REGULATORY COMPLIANCE.

Equipment stoppage during planned operation. Immediate threat to life or limb. Environmental impact or citation is eminent.

[8] DEPARTMENT SHUTDOWN

Work that is not critical enough to require an immediate shutdown but must be performed only during a planned shutdown due to the job content or production schedules.

[7] PREVENTIVE REPAIR

Repairs that are identified and performed to avoid breakdowns, such as a scheduled component replacement prior to the end of its identified lifecycle.

[6] ROUTINE MAINTENANCE

Routine work, such as cleaning and inspecting and asset to proactively identify corrective maintenance needs.

[5] PRODUCTION IMPROVEMENT, QUALITY CONTROL

A malfunction that does not result in a line shutdown but causes product/quality problems.

[4] COST REDUCTION
Work that results in operational changes which will reduce unit costs. Replacement of a defective component with a different component to eliminate or reduce repetitive repairs.

[3] CORRECTIVE MAINTENANCE, SPARES
Fabrication of spare parts or work on spare units. Any corrective maintenance work to eliminate or reduce repetitive work.

[2] PRODUCTION AND SANITATION SERVICE
Fixed assignments or standing work orders.

[1] HOUSEKEEPING
Cleaning of all maintenance areas.

Table 1 - RIME Index Chart

RIME INDEX CHART	SAFETY, BREAKDOWN, REGULATORY COMPLIANCE	DEPARTMENT SHUTDOWN	PREVENTIVE REPAIR	ROUTINE MAINTENANCE	PRODUCTION IMPROVEMENT, QUALITY CONTROL	COST REDUCTION	CORRECTIVE MAINTENANCE, SPARES	PRODUCTION AND SANITATION SERVICE	HOUSEKEEPING
UTILITIES	81	72	63	54	45	36	27	18	9
PROCESS EQUIPMENT (NOT SPARED)	72	64	56	48	40	32	24	16	8
PROCESS EQUIPMENT (SPARED)	63	56	49	42	35	28	21	14	7
SUPPOPRT EQUIPMENT (NOT SPARED)	54	48	42	36	30	24	18	12	6
SUPPORT EQUIPMENT (SPARED)	45	40	35	30	25	20	15	10	5
MATERIAL HANDLING EQUIPMENT	36	32	28	24	20	16	12	8	4
PRODUCTION FACILITIES	27	24	21	18	15	12	9	6	3
PRODUCTION EQUIPMENT (NOT IN USE)	18	16	14	12	10	8	6	4	2
BUILDINGS & GROUNDS	9	8	7	6	5	4	3	2	1

Trend Analysis

Once both production and maintenance management are accustomed to and have confidence in the priority system as a decision tool for sequencing maintenance work. The priority system can assume further responsibility as an analysis tool in evaluating maintenance trends and indices.

To use RIME Index values to evaluate maintenance trends over a period of time, it is necessary to be able to talk of an average RIME value at any specific point in time. The RIME value of a maintenance job is indicative of the importance, or priority, the job should hold, but is independent of the man-hours required to complete the job. Consequently, if RIME indices are to be averaged, their average must be weighted by the associated man-hours of each job.

Table 2 - Weighted RIME Index

JOB	RIME INDEX	X	MAN-HOURS	=	RIME HOURS
A	49	X	24	=	1176
B	54	X	18	=	972
C	90	X	34	=	3060
D	72	X	18	=	1296
		TOTAL	94		6504

$$Average\ RIME = \frac{Total\ RIME\ ManHours}{Total\ ManHours}$$

$$Average\ RIME = \frac{6504}{94}$$

$$Average\ RIME = 69.2$$

Graphical comparison of the average RIME trends, whether for an individual area, department, machine, craft, or any composite basis, can be used for comparing and evaluating trends. RIME values can be used for evaluating the level of maintenance service and give partial answers to such questions as:

- Are we doing more or less relatively important work than previous periods?

- Should overtime be increased/decreased based on the size and importance of the backlog?
- Should Department A receive more preventive or corrective maintenance based on the importance of the jobs requested and completed?
- Are we emphasizing the right jobs based on departmental backlog and completed job history?
- Should our manning be redistributed from Area 1 to Area 2 based on comparative backlog size and importance?
- Is our preventive and corrective maintenance intensity based on the trend of the backlog and its importance?

The analyses of RIME for "Completed" work orders and "Backlog" work orders can be approached from two perspectives: current trends, and long-range trends. The long-range trend analysis is the more meaningful of the comparisons, which examines two basic criteria:

- The "Completed" average RIME line should be higher than the "Backlog" average RIME line.
- The slope of the "Backlog" average RIME line should be either flat or negative relative to the slope of the "Completed" average RIME line.   Satisfaction of the above criteria will indicate the following:

  - The higher priority jobs are properly scheduled first.
  - The jobs being added to the backlog are of an equal or lower average RIME than the jobs that are being worked.
  - Stabilization of these criteria will lead to an operating environment wherein manpower requirements can be decreased.

Accepting that the relative position of the "Completed" average RIME line and the "Backlog" average RIME line meet the established criteria, it is then possible to analyze the current level and slope of the "Completed" average RIME.

Table 3 - RIME Trend Decisions

AVERAGE RIME INDEX LEVEL	SLOPE of TREND LINE	TREND ANALYSIS INTERPRETATIONS
HIGH	UP	Dangerously high level of emergency and breakdown work occurring. Increase maintenance manpower levels with emphasis on Preventive Maintenance (PM) and Corrective Maintenance (CM) during non-operating periods. The action should be considered until the level has been reduced to medium.
HIGH	FLAT or DOWN	The level of emergency and breakdown work is too high. Continued emphasis on PM/CM and increased manpower levels should be considered.
MEDIUM	UP	The overall level indicates a satisfactory mix of work, while the upward slope indicates an increase in emergency or PM jobs. This cannot be determined without specific data. Overtime should be used to control the level of backlog hours as appropriate.
MEDIUM	FLAT	Current mix of work is good. With the slope being flat, the only concern is the size of the backlog.
LOW	DOWN	The current level indicates a good mix of work; however, the downward slope indicates less meaningful work is being requested and completed, a status that indicates an upcoming potential for manpower reduction.
LOW	UP, FLAT, or DOWN	The current level indicates an improper mix of Low Priority jobs being worked, for one of the following reasons: • Improper scheduling of priority work. Analyze backlog trend line for level and slope. • Excess manpower is indicated by apparent ability to accommodate all work requested regardless of priority. Manpower levels should be analyzed.

Work Planning

Once the Backlog has been sorted by work priority, the maintenance work control mega-level process continues with the selection of work for planning. At this stage, the work order status changes to "In-Planning" to reflect that the job is progressing within the work order system. The work order remains "In-Planning" until which time the job package is "Ready" to be scheduled, meaning that all labor and material resources are available to schedule and complete the task.

Whenever maintenance needs arise, planning takes place. The questions are, "who's doing the planning" and "how effective was the plan?" In a reactive maintenance environment, where most of the work is performed after a breakdown or downtime event has occurred, work planning is performed by the Maintenance Technician, extending the maintenance turnaround time and equipment downtime.

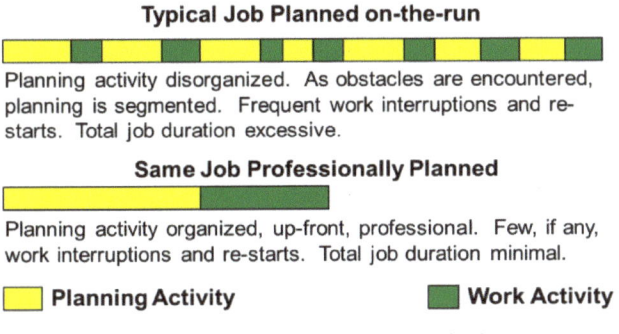

Typical Job Planned on-the-run

Planning activity disorganized. As obstacles are encountered, planning is segmented. Frequent work interruptions and re-starts. Total job duration excessive.

Same Job Professionally Planned

Planning activity organized, up-front, professional. Few, if any, work interruptions and re-starts. Total job duration minimal.

Planning Activity   Work Activity

Figure 4 - Reactive vs. Proactive Work Planning

Within the proactive maintenance workflow process, work planning is formalized through a dedicated "Planner" who is responsible for analyzing the scope of work, preparing the job plan, and coordinating the parts, materials, special tools and permits required to complete the work efficiently. Typically, formal work planning reduces the maintenance turnaround time by 30% to 60% by reducing the frequency of job interruptions.

Labor Utilization Impact of Work Planning

The difference between reactive and proactive workflow management is the volume of work that is planned and scheduled in advance of the work being completed by Maintenance Technicians. As a comparison, the Direct Labor Utilization Chart (Table 4) illustrates the impact that formal work planning has on maintenance labor utilization. "Direct" labor refers to the amount of time Technicians spend inspecting, detecting, cleaning, lubricating or repairing an asset. This is often known as "wrench time". "Indirect" labor is the time spent administering the workflow process, including planning the job, or time allocated to authorized breaks, scheduled meetings, shop cleanup, and unauthorized personal time.

Table 4 - Direct Labor Utilization Chart

Task	Reactive	Proactive
Receiving Instructions	5%	3%
Obtaining Parts, Materials or Tools	12%	5%
Travel To/From the Job Site	15%	10%
Idle at Job Site	5%	1%
Job Interruptions (Scope or Job Change)	8%	4%
Other Indirect Time	20%	12%
Subtotal	65%	35%
Direct Labor Utilization	**35%**	**65%**

The Planner is the person primarily responsible for managing the flow of work through the system. To do so, he uses a system of work order Status Codes to define where each job is in the work order lifecycle and to monitor direct and indirect labor utilization.

Table 5 - Work Order Status Codes

Work Order Status Codes			
Numeric	Alpha	Description	Role
1	A	Approved	Supervisor
2	W-P	Waiting to be planned	Supervisor
3	W-E	Waiting for engineering support	Planner
4	W-M	Waiting for management approval	Planner
5	DF	Deferred - pending funding	Supervisor
6	PO	Waiting for PO to be issued	Planner
7	DFP	Waiting for materials (down for parts)	Planner
8	DFT	Waiting for tools (down for tools)	Planner
9	W-W	Waiting for weekend downtime	Planner
10	W-A	Waiting for equipment availability	Planner
11	KIT	Work order kitted	Storeroom
12	R-S	Ready to schedule	Planner
13	S	Scheduled	Planner
14	S-D	Scheduled - deferred	Supervisor
15	I	Work in progress	Supervisor
16	I-D	Work in progress - deferred	Supervisor
17	C-M	Complete - material reconciliation pending	Planner
18	C-R	Complete - rebuild order pending	Planner
19	C-E	Complete - engineering documentation pending	Planner
20	C	Closed	Planner

Level 1	Level 2	Level 3	Level 4	Level 5
Work order status is not used; all Work Orders are entered as same status (i.e. "Released", "Approved").	Work order status codes are not understood; Work order management is informal; Planner completes "mass close" to purge completed work orders.	Workflow process is formally document; Planners understand and apply statuses properly; No formal reporting.	Workflow process is bound by Work Order Status in CMMS; Evidence of consistent compliance exists; No significantly aged work orders.	Level 4 plus Management /Engineering monitors workflow cycle time; Work order status is used to identify turnaround time improvements.

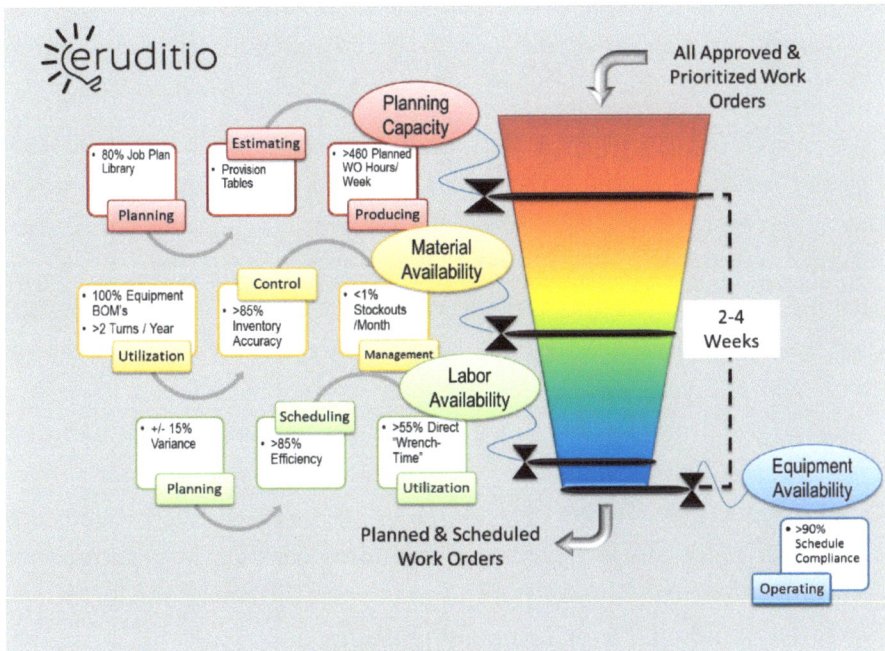

Figure 5 - Backlog Funnel Work Management Model

Planning Performance

To support "Backlog Management" (Figure 5) it is important to discuss the level of performance within the Work Planning process that is required to ensure the backlog of work is maintained within acceptable

27

levels, 4-6 weeks for a "Total Backlog", and 2-4 weeks for the "Ready Backlog".

The first action within the Work Planning process is to select the appropriate work orders from the backlog. In any work prioritization model, this workload is made up of those work orders that have a lead-time of greater than seven days, at a minimum. Ideally, however, it is recommended that work planning begin as early as fourteen days prior to the expected delivery, or a lead-time. Obviously, using RIME, it is easy to recognize that only those work orders that have an index number of less than 72 will be selected for formal planning. With this said, however, we should not assume that higher priority work orders do not require work planning. In fact, because these activities are "critical" to production, process, or people, efforts should be made to ensure that the maintenance activity is efficient, effective, and performed at a high standard of service quality.

The performance model for Planners is based on a good maintenance practice span-of-control of one Planner to twenty Technicians (1:20). The method of monitoring is determined by the number of work order hours planned in relation to the average direct labor hours available, or 520 man-hours of planned work from each Planner with and average direct labor utilization of 65%. If the span-of-control is reduced, then the expected performance is reduced accordingly. A common mistake that organizations make is in the way they define the Planner's role. Organizations that define the Planner as an area-based resource are often less successful at managing the total backlog than those that centralize their planning function to ensure maintenance resources are expensed based on the highest _plant priorities_.

Job Planning

Maintenance Job Planning/Estimating Worksheet						
Planner: John Doe		WO#: 2012-AFE-000125			Date:	8/15/2012
Job Contact: Fred Doe		Location: Cooling Tower No.2 Mechanical Room			Asset ID#:	PMP6543
Job Description: (list acceptance testing criteria) Replace pump impeller in order to restore pump functional capacity to 1300 GPM. Replace shaft seal and pump casing gasket to prevent water leakage during operation. Inspect pump shaft bearings for signs of accelerated wear, including metal debris, discoloration, and visible cracks or surface defects, and replace as required. Ensure pump-motor shaft is aligned to eliminate defects caused by excessive vibration.						
Job Scope						
Job Plan Sequence	Task Description	Craft/Skill	Task Duration	Number of Resources Required	Total Estimated Labor Hours	Total Actual Labor Hours
10	Inspect pump casing for corrosion	MECH	0.25	1	0.25	
20	Disconnect pump shaft coupling	MECH	0.25	1	0.25	
30	Remove upper casing	MECH	0.5	2	1	
40	Remove rotating element	MECH	1	2	2	
50	Sandblast upper and lower pump casings	MECH	1	1	1	
60	Inspect upper and lower casings for cracks	MECH	0.25	1	0.25	
70	Weld casing cracks	WELD	1	1	1	
80	Clean upper and lower casing	MECH	0.5	1	0.5	
90	Inspect pump shaft bearings	MECH	0.25	1	0.25	
100	Install new rotating element	MECH	1	2	2	
110	Install new casing flange gasket	MECH	0.25	1	0.25	
120	Install upper casing	MECH	0.25	2	0.5	
130	Verify shaft alignment	VIBE	0.5	1	0.5	
140	Record system parameters	MECH	0.25	1	0.25	
150	Clean area around pump	MECH	0.5	1	0.5	
160					0	
Total Job Plan Duration:		7.75	Total Backlog Labor Hours:		10.5	0

Figure 6 - Job Plan Template

Creating the "Job Plan" is more than opening a work order and assigning labor. The "plan" must break the scope of work into manageable actions or tasks (Figure 6), with the labor and material requirements documented for each task.

Some job plans may require the Planner to develop another level of detail based on the complexity of the job. Standard Work Instructions (SWI) or formal work procedures may be required for jobs that include:

- Multiple craft skill requirements,
- Inter-departmental coordination,
- Durations that extend past normal shift schedules,
- Critical tasks that, if performed incorrectly, could result in injury or equipment damage, or
- Infrequently performed tasks.

Job planning also requires the Planner to document potential hazards exposed to the Technician during work execution. Permits and Lockout-Tagout (LOTO) procedures should also be included in the job plan where they apply.

Work Scheduling

The Scheduling Team

A scheduling team shall be established with appropriate membership from the following roles:

- Maintenance Planners
- Maintenance Supervision
- Operations Supervision
- Engineering Personnel
- Materials Management Personnel

Specific expectations for the scheduling team shall be established and communicated throughout the organization. An emphasis on the active participation of operations personnel in the development of the work schedule shall be given a priority in the scheduling process.

A schedule coordinator should be selected from within the team, who will have the specific responsibility of leading the scheduling meetings, documenting the agreed upon schedule, and publishing the schedule in the agreed upon manner. In many organizations the Maintenance Planner also serves as the schedule coordinator; however, it is common to separate the work planning and work scheduling role responsibilities to focus Maintenance Planners on developing an effective library of pre-planned jobs.

Leaders from the organization (Superintendents, General Managers, etc.) should sponsor and support the scheduling process. This is demonstrated through periodic attendance at scheduling meetings, reviewing published schedules, and monitoring schedule compliance.

Weekly Scheduling Process

A standard process defining the weekly scheduling activities to be performed by the organization must be documented and consistently followed. Whenever possible, this process should leverage standard location and times for these activities to facilitate consistent execution.

At a minimum, the weekly scheduling process would contain the following elements:

- Review of the previous week's schedule compliance,
- Cut-off date for determining resource availability and non-emergency add-on work,
- Scheduling meeting agenda, date/time, and format,
- Contractor schedule communication protocol,
- Communicating maintenance material needs to facilitate kitting and staging activities, and
- Schedule publishing and communication protocol.

Scheduling Time Horizons

To allow adequate time to prepare for the work to be executed, it is in the organization's best interest to produce schedule projections into the future. The greater the scheduling projections, the more time the organization has to coordinate the requirements identified.

Figure 7 illustrates a progressive 4-week scheduling time horizon. In the beginning stages of developing a scheduling process, the organization should focus on the 1-week time horizon and attempt to master the skills required to maximize schedule compliance in this time window. As the organization matures, the time horizon should be extended to include a 2-week, and eventually a 4-week time horizon.

Within each time horizon, work scheduling must balance three types of work, Preventive Maintenance (PM), Corrective Maintenance (CM), and Emergent Maintenance (EM) – labor allocation required to respond to unplanned or urgent maintenance requests, like a breakdown. The preferred model for a 1-week time horizon is 30% PM, 60% CM and 10% of the net available labor _assigned_ to respond to emergent or urgent requests.

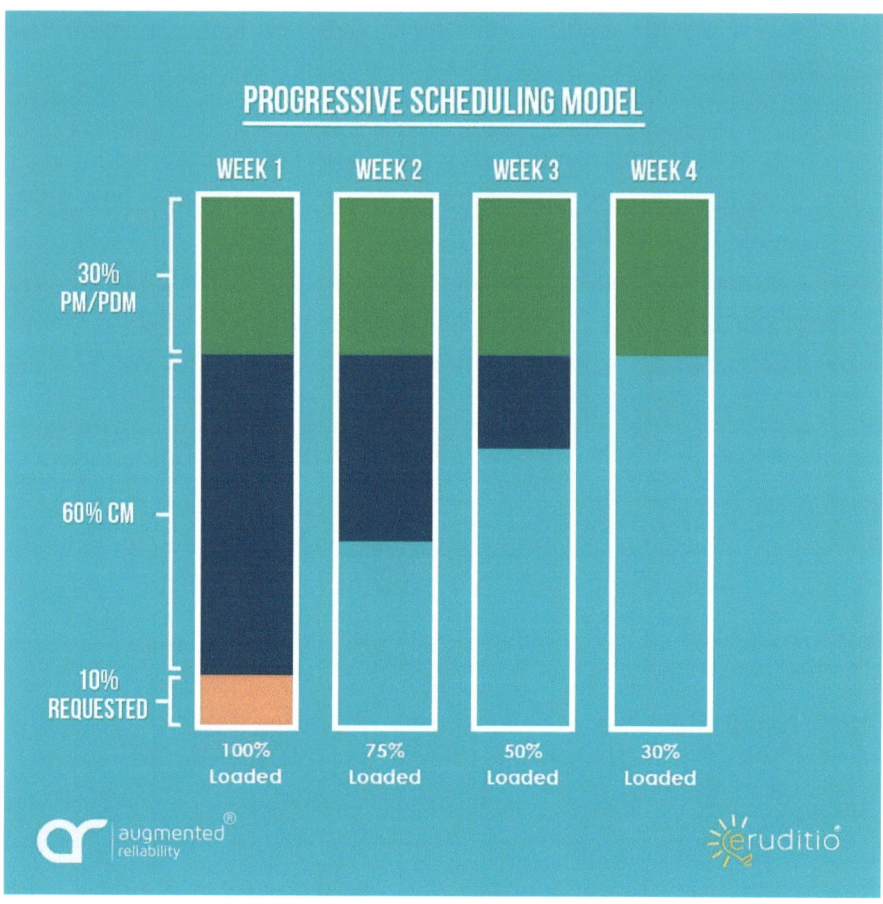

Figure 7 - Progressive Scheduling Model

Scheduling Meeting

A scheduling meeting should be held on a weekly basis with mandatory attendance by the scheduling team. A standard scheduling meeting agenda would be established and followed each week. The prioritized "Ready Backlog" will be the driving force behind the development of the weekly schedule. Only those work orders that have all of the required materials and labor available to complete the scope of work should be included in the scheduling discussion.

Resource Utilization

The goal of the weekly scheduling activity is to maximize the use of available resources (Technicians). The amount of time scheduled for each available resource must be tracked and every effort made to maximize the use of these resources, with continuous improvement activities being leveraged based on the schedule compliance measures.

No time shall be held back to allow for possible emergency work, and 100% of available resources should be accounted for on the weekly schedule. However, a robust maintenance work control process should accommodate for any emergencies that occur. This is accomplished by balancing labor distribution across the three types of maintenance: Preventive (PM), Corrective (CM) and Emergent (EM) as shown in Figure 8.

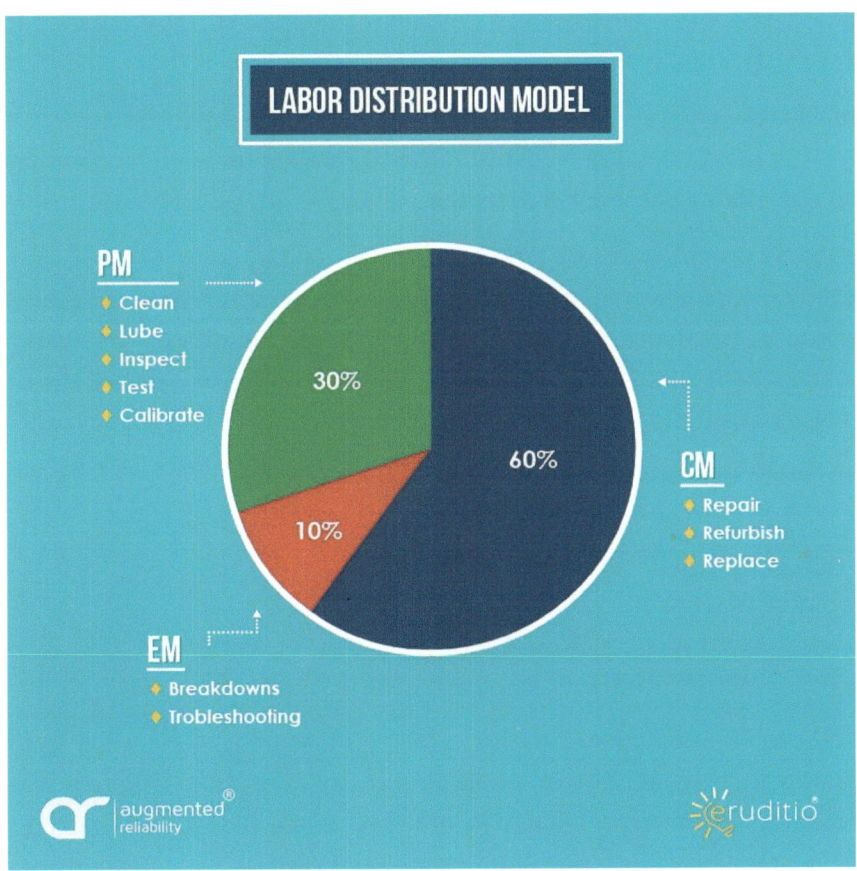

Figure 8 - Labor Distribution Model

Within the weekly schedule, maintenance labor should be distributed to ensure that all work can be completed, including responding to urgent or emergent requests without negatively impacting the scheduled PM and CM volume of work. Maintenance labor resources should be _assigned_ to each work order on the weekly schedule.

Publishing and Communicating the Schedule

The weekly schedule must be physically published, visible to all stakeholders, and actively communicated to the personnel who will eventually be affected by that schedule (crews, operations, stores, etc.). Every effort should be made to publish and communicate the schedule earlier in the previous week to allow the workforce to review and prepare for the following week's schedule.

Emailing or posting the schedule on a networked drive should not be considered an adequate substitute for physically publishing and actively communicating the weekly schedule.

Schedule Compliance

Schedule compliance must also be measured and tracked. In addition to schedule compliance, the detractors to schedule compliance, also known as "Schedule Breakers", should be documented and reviewed on a weekly basis by Senior Management.

Documenting the detractors from schedule compliance should include:

- Reasons for breaking the schedule,
- Add-on work,
- Deferred work, and
- Jobs that exceeded the estimated time and resources.

The Schedule Compliance metric is first a measure of adherence to the maintenance schedule expressed as a percent of total schedule work order hours, and second as a measure of resource allocation as a percent of the "Net Available Labor Hours" per week. Net Available Labor Hours is derived by subtracting the indirect labor hours resulting from breaks, reallocation of labor (i.e. special projects), absenteeism, PTO and other hours that prevent maintenance resources from being scheduled for work orders in the "Ready Backlog".

$$Schedule\ Compliance = \frac{Scheduled\ Work\ Order\ Hours\ Completed}{Scheduled\ Work\ Order\ Hours}$$

$$Schedule\ Efficiency = \frac{Scheduled\ Work\ Order\ Hours\ Completed}{Net\ Available\ Labor\ Hours\ per\ Week}$$

Table 6 - Daily Management Metrics

Metric	Definition	Target	Reported by	Action Plan
Schedule Compliance	Actual work orders completed as scheduled	85%	Scheduler	75% - Review noted "non-compliance" at Daily Management Meeting
Planning Accuracy	Actual labor hours compared to estimated labor hours	± 15% Variance	Planner	Initiate formal review and audit of work orders
Planned Work Percentage	Volume of work planned vs. total work executed	90%	Planner	Review planned work versus total work executed
MRO Inventory Accuracy	Actual material level on-hand compared to inventory level noted in EAMS	98%	Storeroom Supervisor	Reconcile requisitions against "issued" materials

Work Delivery Review & Feedback

Work Delivery is the next step in the maintenance work control mega-level process and can be broken down into two distinct steps, the first of which is Work Delivery & Feedback. The Work Delivery & Feedback process begins with the communication of the weekly schedule, which has assigned the work order or job package to a specific work delivery crew 2-weeks prior to the scheduled start date and time, ideally. The Maintenance Supervisor than accepts the assignment of work by formally reviewing the job package for completeness and accuracy, ensuring that the job can be successfully and efficiently executed as planned and scheduled. This also provides the work delivery crew with an opportunity to review the job package and forward any enquiries to the appropriate "Planner" or Maintenance Engineer for clarification of procedure, maintenance standards, and commissioning or test parameters. Additionally, as the required tools and materials have been confirmed prior to scheduling of the work order, the Work Delivery Review & Feedback process allows the Technician to review the kit prior to executing the work to verify the materials identified and provided meet the equipment specification, providing a final opportunity to remove inefficiency from the execution of maintenance.

Work Delivery Execution & Feedback

The second step is Work Delivery Execution & Feedback, which outlines the process by which the job package or work order is completed, beginning with the delivery of materials to the work site. Although the delivery of kitted materials to the work site is a Materials Management work control process, it is important to identify where the "Delivery" process intersects with the execution of maintenance. In most organizations, kitted materials are delivered to the work site or a secured staging area 24-hours prior to the scheduled start date and time.

The Work Delivery Execution & Feedback process continues until which time the scope-of-work is fulfilled, the equipment is commissioned and accepted by the asset owner or Production representative, and equipment, work, and failure history data are collected on the work order. This aspect of the maintenance work control process again requires integration with Reliability Management processes to ensure that the necessary data is clearly identified within the job package or work order.

Once the criteria for completing the job package or work order have been satisfied, the work order transitions to the "Complete" status, indicating that the scope-of-work has been satisfied and the equipment has been returned to normal operating condition. In the event where the task or work order is a subset or "child" of another work order, only the completed task or "child" work order is progressed, allowing the "parent" work order to remain open until all tasks have been completed. This practice allows for meaningful analysis by task, asset or crew, and ensures that schedule compliance is "live", providing real-time information to managers as to the current plant condition.

The final step within the Work Delivery Execution & Feedback process is to initiate any follow-up corrective actions resulting from the completed work.   This is especially important when completing Preventive Maintenance (PM) tasks as their effectiveness is measured in part by the number of Corrective Maintenance (CM) work orders generated in relation to the number times the PM work order has been completed. Known as the "6:1 Rule", a minimum of one (1) CM work order should be generated for every six (6) times the PM work order is completed.

The 6:1 Rule Explained

The "6:1 Rule" is more than a rule of thumb. It is a statistical evaluation of Preventive Maintenance effectiveness and provides an early warning indication of too little detail in the PM job plan, training deficiencies, or the fact that the PM itself is not appropriately designed for the types of failure mode characteristics associated with the asset.

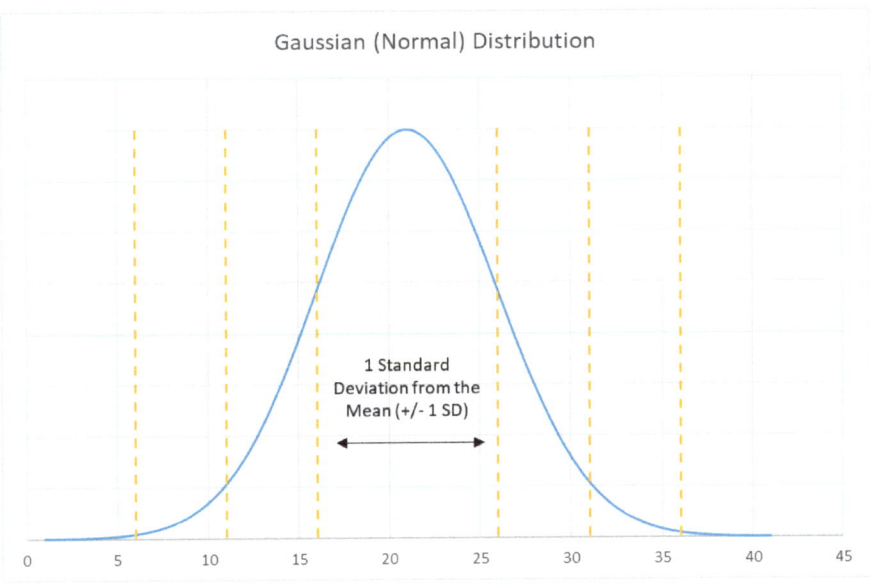

Figure 9 - Normal Distribution Curve

Figure 9 demonstrates the normal distribution of random failures, with three standard deviations to the left and right of the peek of the curve,

37

the "Mean" – central tendency or average of the data. Within the three standard deviations exists 99.73% of all the reported failures over the analyzed period. If we assume that the Preventive Maintenance (PM) work orders are scheduled at a frequency equal to one standard deviation from the mean, then we should find evidence of a failure mode once for every six times the PM is completed. "6" in the 6:1 Rule refers to the six PM inspection opportunities across the normal distribution of failures.

Work Closeout

The last operation in the maintenance work control mega-level process is the Work Closeout process, which facilitates the formal integration of Maintenance and Reliability Management work control processes through the development of work history records.

Often time's organizations omit the Work Closeout process, as it appears to be a redundant, thus a wasteful mode of operation. In fact, however, Work Closeout is one of the most important aspects of the maintenance work control process as it is the only means of creating "history" within the Computerized Maintenance Management System (CMMS), not to be confused with the collection of data for history, as in the fore-mentioned process.

The Work Closeout process was a necessary development resulting from the nature of relationship-database configurations used in most, if not all, Computerized Maintenance Management Systems today. Because data within the CMMS is based on key elements which link, or relate data to each other, historical records are not compiled until which time the data becomes secure or "locked". To explain this more simply, the work order data is not reported against the asset until the work order is transitioned to the "Closed" status, affectively locking the work order and the data to protect its integrity. The most common problem with this maintenance management practice is that it may conflict with your organizations accounting practices, especially in respect to contract services or direct-charge purchases that may be allowed to remain open and unreconciled for a predetermined period. That predetermined period in which the purchase order can remain open against the work order prevents the work order from being "Closed". What ensues are inaccurate work histories. Failing to understand this aspect of work control will cause Maintenance Engineers and Management to be misguided by inaccurate data, enabling them to potentially react to the wrong improvement opportunities.

Level 1	Level 2	Level 3	Level 4	Level 5
Failure or coordination delay codes aren't identified on work orders; Comments are rarely provided by Technicians.	Work orders are closed with Spare Parts information; No actual labor reporting; No delay codes.	Consistent Closeout with Spare Parts and Actual Labor information; Delays >60 minutes are noted; Written feedback provided on <50% of work orders.	Level 3 plus Failure Codes identify Part, Problem, Cause and/or Remedy; All coordination delays >30 minutes are recorded; Written feedback on >50% of work orders.	Level 4 plus work order closeout is completed daily; Trend analysis enables opportunities to reduce Mean Time To Repair.

THE MAINTENANCE ORGANIZATIONAL STRUCTURE

One of the core principles of proactive Maintenance work management is that the three functions of maintenance management must be provided for within the organizational design: Supervision of work execution, planned work preparation, and maintenance engineering. In this chapter we will define the roles and responsibilities of each function and examine the organizational design requirements based on Maintenance Technician full time equivalents (FTE) or "headcount".

Maintenance Supervisors

Maintenance Supervisors control the safety, quality and on-time completion of work execution. Supervisors should be on the shop floor, available to Maintenance Technicians, and observe the work being performed. Supervisors should strive to visit each Maintenance Technician a minimum of twice a shift or workday. During major shutdowns or critical, complex jobs, Supervisors should visit the work site at the beginning, middle and end of the job.

Beginning the Job

- Confirms that the proper number of Technicians are assigned to each job.
- Instructs Technicians on what is to be done and, as appropriate, the methods to be used. Reviews safety requirements and communicates potential hazards.
- Checks that necessary tools, materials and parts are available prior to beginning the job.
- Walks through the job, physically or mentally, to ensure the job can be completed ergonomically and safely.

During the Job

- Keeps the job moving.
- Regularly accounts for his Technicians and adjusts manpower as needed for an on-time completion of the job.
- Stays in close contact with the progress of the job and eliminates bottlenecks.
- Immediately informs the Planner if a job cannot be completed as planned, and reallocates labor as required.

Completing the Job

- Verifies that all required work has been completed.
- Sees that the shop or job site is cleaned and left in good order.
- Obtains sign off by representative of the requesting organization and marks the work order "Complete" in the work order system.
- Returns completed work orders to the Planner, and records schedule breakers or notes to improve work performance.
- Returns prints, sketches, drawings and other job documents to the Planner for proper filing to reduce preparatory efforts next time a similar job is performed.
- If "as installed" conditions have been changed, ensures that prints and drawings have been red-lined for engineering revision.

Maintenance Planner (Planner/Scheduler)

The role of the Maintenance Planner is to improve work force productivity and work quality by anticipating and eliminating potential delays through planning and coordination of manpower, parts and materials, and asset availability.

Reporting to the Maintenance Superintendent and a liaison between Operation and Maintenance, he is responsible for planning, scheduling and coordination of all plannable maintenance work performed on the plant site. Through supervision of the clerical resources, he is also responsible for maintenance records, work histories and equipment data that is essential for meaningful performance analysis and failure reporting.

Initial Job Screening

The Planner receives all requests for maintenance work, except those that must be performed on the same day as requested (i.e. Emergent (EM)). Emergent requests are handled directly by the Maintenance Supervisor without the benefit of formal planning.

- Reviews and screens each request for completeness, accuracy and clarity of scope.
- Develops preliminary estimated if required to obtain approval.
- Obtains Engineering approval for all equipment alteration and modification requests.

Analysis of Job Requirements

- Consults with the requester, the Maintenance Supervisor or the Maintenance Engineer as needed to accomplish the requested scope of work.
- Visits the job site and analyzes the job in the field (20% time allocation) to determine special requirements, such as scaffolding, fall protection or confined space entry.
- Identifies potential hazards to Technicians during the execution of the job. Prepares all Lockout-Tagout (LOTO) and permitting requirements for the job.
- Evaluates existing master-data Job Plans to determine if the scope of work was previously planned.

Detailed Job Planning

- Breaks the scope of work into actionable tasks:
    - Identifies the required parts, materials and special tools for each task.
    - Identifies the required Craft Code or skill level for each task.
    - Estimates the duration of time (i.e. labor) for each task.
- Creates the Bill of Materials (BOM) for the job plan and submits the Materials Request, electronically or in paper form.
- As required based on job complexity, the Planner writes standard work instructions or work procedures to accomplish critical tasks within the job plan.

- Assembles prints, sketches, drawings and pictures needed to clarify critical tasks.
- Opens child work orders – work orders assigned to the primary or parent work order – as necessary to schedule contractors, vendors and other departmental resources external to the Maintenance organization.

Job Scheduling and Coordination

- Evaluates Maintenance Technician "Net Labor Availability" – payroll hours less indirect commitments.
- Reviews asset availability windows for those job plans requiring the equipment to be shut down.
- Confirms the readiness and availability of kitted materials.
- Updates scheduling time horizons with "Ready Backlog" work orders.
- Performs workload smoothing to balance PM, CM and requested volume of work within the scheduling time horizons.
- Facilitates the Weekly Scheduling Meeting to achieve a consensus between asset owners (i.e. Production and Engineering) and Maintenance Supervisors based on equipment and business needs.
- Publishes the Weekly Schedule, and subsequent Daily Schedules as required by Maintenance Supervisors by crew, Craft Code or physical plant location.

Job Close Out and Follow Up

- Publishes the "Daily Management Metrics" and reviews schedule breakers with Maintenance Superintendent.
- Reviews completed work orders and records work history – actual labor and materials – and Failure Code data.
- Updates master-data Job Plans to improve schedule efficiency and planning accuracy.
- Reconciles open requisitions for maintenance, repair and overhaul spare parts and materials.

- Reconciles open requisitions for services performed by contractors, vendors or other departments external to the Maintenance organization.
- Submits to Engineering any and all requested changes to prints, drawings, or documents pertaining to the installed configuration of plant assets.
- Opens follow-up corrective maintenance (CM) work orders resulting from completed PM or EM work orders.

Maintenance Engineer

The role of the Maintenance Engineer is to enhance service quality and asset reliability by focusing on workflow improvements, optimizing maintenance strategies, and championing process improvement projects.

The Maintenance Engineer plays a critical role in connecting his assigned plant operating center or department with the Maintenance organization, ensuring minimal downtime and failure rates, and optimum maintenance turnaround leading to improved asset utilization.

The Maintenance Engineer also serves as the Maintenance liaison to Engineering and Production departments to help improve asset design for reliability and maintainability.

People & Training

- Keeps current with new equipment and technology and updated maintenance methods. Learns and appreciates the importance of Service Quality, Service Delivery and flawless job execution.
- Coaches Maintenance Supervisors on process improvement techniques and skills.
- Identifies and coordinates craft skills training, and equipment operator training to reduce defects caused by human error.

Process & Equipment

- Competent practitioner of Reliability Centered Maintenance (RCM) techniques and utilizes these skills to optimize equipment maintenance strategies.

- Understands the theory and application of predictive maintenance technologies such as Infrared Thermography, Ultrasonic Testing, Vibration Analysis and Tribology.
- Actively assesses opportunities to incorporated predictive maintenance techniques and technology into current equipment maintenance strategies.
- Competent practitioner of Root Cause Analysis and utilizes these skills to facilitate investigations of non-conformance events.
- Actively uses business systems to analyze the effectiveness of technical solutions and capture knowledge sharing.
- Is a competent practitioner of LEAN Process Optimization practices and utilizes these skills to participate in and lead LEAN continuous improvement projects.
- Competent practitioner of Six Sigma Quality Assurance practices and utilizes these skills to participate in and lead Six Sigma projects with a focus on eliminating asset performance variances.
- Is an active and contributing member to the locations Loss Prevention Team and Quality Assurance Team. Follows up on maintenance related, remedial action plans.
- Is actively involved in evaluating the effectiveness of Standard Maintenance Procedures and Root Cause Analysis triggers.
- Monitors the cost of service delivery (labor and materials) with a focus on reducing the cost per operating hour for his assigned plant operating center.

Quality & HSE

- Assists Management in defining and driving the implementation of the Service Quality Plan within the assigned plant operating center.
- Actively participates in Service Quality Investigations, remedial work planning, and continuous improvement planning.
- Assists supervision with adherence to Service Quality Standards by being actively involved in the auditing and implementation of corrective action plans.

- Supports Management efforts to minimize Quality and HSE risks and promotes respect, understanding and adherence to HSE regulations.
- Continuously strives to optimize the safety and efficiency of Maintenance processes and procedures.

Maintenance Organizational Design

Maintenance organizations come in all shapes and sizes, from small multi-skilled machine shops to large multi-site service organizations. The number of Maintenance staff roles required to support the proactive maintenance workflow process, like Supervisors, Planner and Engineers, is determined by the number of Maintenance Technicians required to sustain a 4-6 week "Total Backlog" of work.

Maintenance staff roles, and their ability to support the maintenance workflow process, should be based on the best practice ratios as defined in Table 7.

Table 7 - Maintenance Span of Control Chart

Staff	Ratio to Technicians
Maintenance Supervisors	1:12
Maintenance Planners	1:20
Maintenance Engineers	1:40
Maintenance Clerks	1:40
Maintenance Training Coordinators	1:70

Maintenance organizations may also differ in terms of the reporting relationship to Senior Management based on the overall organizational design. In those organizations that are Production-centric – staff reports to a Senior Production or Line Manager – Maintenance staff is decentralized and generally reports to Management within a specific production center or operating department. Within a Functional organizational structure, where Maintenance is centralized, Maintenance staff reports directly to the senior Engineering Manager or Maintenance Manager.

Although very supportive of today's high-performance culture, decentralized structures create higher levels of risk due to the competing priorities between asset utilization and asset health management. Centralizing your Maintenance organization will provide more consistent direction and communication to alleviate resistance due to miscommunication and conflicting asset management priorities. A centralized structure is recommended for those organizations that are transitioning from "reactive" maintenance with their sights set on managing the proactive maintenance workflow process. Use the "Maintenance Maturity Assessment" (Table 8) to evaluate your workflow process and identify areas of opportunity to improve the performance of your asset management plans and practices.

Table 8 - Maintenance Maturity Assessment

	Elements	Level 3	Level 4	Level 5
Work Execution Management	**Work Control**	Work control process is formally documented, and standards of practice exist for each step, task, or function of work control.	Level 3 plus stakeholders and resources have been trained to execute steps or tasks in accordance with the work control process.	Level 4 plus formal procedures exist for measurement and monitoring of work control efficiency.
	Work Order Usage	Work orders are consistently generated for proactive work. Reactive work is undocumented.	All work is performed against a work order. 100% of all labor and material expenditures are recorded against a work order.	Level 4 plus work and failure histories are used to optimize Asset Management plans.
	Work Order Prioritization	Priorities are driven by either asset criticality, defect severity, or work order type without simultaneous consideration. A formal process exists but is not consistently executed.	Priorities are driven by both asset criticality and defect severity; work order type is not considered. Formal process documented and consistently executed.	Priorities are driven by asset criticality, defect severity, and work order type simultaneously. Formal process documented and consistently executed.
	Work Procedures/Job Plans	Job plans exist for multi-craft jobs on critical machines but format is poor, resulting in inconsistent execution.	Job plans exist for most jobs on critical machines and follow a best practices format with consistent execution.	Level 4 plus a job plan library by task exists for 90% of work. Craft personnel consistently provide input to improve MTTR.
	Backlog Management	Work orders have estimated hours assigned and backlog is known in total number of weeks.	Level 3 plus "Ready to Schedule" backlog is easily identifiable and decisions are made to balance service crews using backlog calculation.	Level 4 plus Top Management closely monitors backlog trends to determine proper staffing and outsourcing needs.
	Resource Management	Scheduling is based on availability of resources, but only scheduling to 80%. Direct labor utilization is < 50%.	Scheduling > 90% of labor availability. Direct labor utilization is 55%-65%.	100% of available labor hours are scheduled; formal process is in place to address non-compliances (e.g. "schedule breakers"). Direct labor utilization is > 65%.
Work Execution Management (continued)	**Equipment Bill of Materials (BOM)**	BOMs exist for > 60% of all <u>maintainable</u> assets.	Level 3 plus inventory accuracy is > 98%.	Level 4 plus formal procedures exist to add, revise, or delete items for the BOM.
	MRO Inventory Management	> 60% of SKUs are managed by the use of ABC classifications. Inventory turns are 2-3 times per year.	Level 3 plus a routine cycle counting process exists based on ABC classifications.	Level 4 plus inventory levels and turns are routinely optimized based on Asset Management plan requirements.
	Material Kitting	Material kitting and staging occurs for > 65% of outage jobs but ad hoc for weekly schedules.	Material kitting and staging occurs for > 65% of all work. Storeroom personnel receive pick-lists and assemble the kits. Kits are identified by work order and level of completeness.	Level 4 plus kits are kept in a secure area, verified for accuracy against the work order, and are easily identified by requestor.

www.ingramcontent.com/pod-product-compliance
Lightning Source LLC
Chambersburg PA
CBHW041111180526
45172CB00001B/209

* 9 7 8 1 0 7 6 6 9 1 9 9 6 *